The Dragon-
School Bus

A Courageous Fight to Clean the Air

Louie Dias

Illustrated by Evan Winston

Published by Louie Dias Books

Inquiries should be addressed to Louie Dias Books, please email louiediasbooks@gmail.com.
Sacramento, California

To learn more, visit the website at louiediasbooks.wordpress.com

Summary:
Henry is a creative boy who loves dragon-style battle cards. But asthma attacks are keeping him out of school. His caring school bus driver sees that her polluting diesel bus is harming the air. Together they'll courageously fight for a new clean electric bus, and make a difference for their community. But you have to believe...

Text Fonts: Sawarabi Mincho and Sorts Mill Goudy
Illustrations: Photoshop

ISBN -
Hardcover: 978-1-7371451-1-0
Paperback: 978-1-7371451-0-3

Library of Congress Control Number: 2021919193

[1. Electric Vehicles - Fiction. 2. Asthma - Fiction. 3. Climate Change - Fiction]

Dedication

To those who have suffered from smoke and other pollutants.

Too many precious lives have been taken because we couldn't imagine a better way. Many are now answering the call for environmental justice, may this book help win the fight.

Ignorance can no longer be an excuse...

Special thanks to a cast of editors that helped bring this story home: Brooke, Aileen, Susannah, Yazmine, and the 2nd grader that said "There should be a vote."

Henry waited for the bus, holding his new dragon card. The glass doors opened with a "whoosh."

He raced up the steps to the driver's seat.

"Miss Leona, check out my new battle card! It's a *Chrystal Eyes Green Dragon*. I can summon it with two spell cards. I'll CRUSH my opponent!"

"Like a grape, I'm sure," she laughed.

Miss Leona saw his hospital wristband. "You been OK? I've missed you."

“I was at the Emergency Room,” he said softly. “I get asthma attacks. It’s hard to breathe. Like a *Red-Eyed Plasma Dragon* is sitting on my chest. It’s worse when I run hard, or breath dirty air.”

Suddenly he began coughing and wheezing.

Henry searched his backpack, and found a small plastic tube. He squeezed and breathed in the medicine.

"This inhaler stops the attack on my body. I don't want to miss any more school."

"My son had asthma attacks too," she said. Just like back then, she felt powerless to help.

The bus chugged down city streets. Thick black smoke billowed out the tailpipe.
Henry grimaced - sometimes he could taste and smell the diesel fumes.

"I'm sorry about this fire-breathing dragon," said Miss Leona. "I've been breathing his smoke for 20 years."

Henry tilted his head. "Dragons might be big and fierce, but people are smarter. We can defeat them if we use science and work together. But you have to believe!"

That day Henry visited the school library. The books he checked out barely fit in his backpack.

The next morning, Henry climbed the steps slowly. "What's wrong?" asked Miss Leona.

"I woke up in an asthma attack," he said. "My lungs felt plugged up. It was hard going back to sleep."

"That sounds terrible."

"It's OK," he said. "My cards kept me company. And I read about how we can clean the air."

"Oh? What can we do?"

"Electric buses don't burn diesel and gas, so there's no smoke or chemical smell," he explained.

"Electric? Like a remote-controlled car?"

"Yeah, but it's batteries are way bigger!" he said. "And electric cars put energy back into the batteries, to use again."

"Hmmm. Like recycling something, instead of throwing it away," she said.

"Yep. And electric cars are quick. Like my *Crescent Fang Moon Dragon!*" he exclaimed, arms out in battle pose.

Henry looked down. "Miss Leona, will you try to get a new bus? I like our drives, and I want to get healthy."

"I want that too," she said.

She returned to the bus yard, and explained to her boss why an electric bus would help.

"Sounds great," he said. "And expensive."

"My old bus needs to be replaced anyway. And we'd never have to buy fuel again. That would save thousands."

He shook his head. “Let’s just wait...”

Miss Leona stood tall. “Wait for what? More sick kids and dirty skies? We can’t keep waiting to do what’s right.”

All he could say was, “You’re right.”

The principal agreed, because kids with asthma miss recess and sports.

"There's one more big step," Miss Leona told Henry. "We have to convince the school board. Will you come with me?"

Henry nodded. "I'll summon my dragon strength."

At the meeting, the leader asked Henry why he came.

"Electric cars are quicker, cleaner, and safer," he explained.

"Anyone could tell us those facts, Henry. Why is this important to YOU?"

Henry remembered Miss Leona's son. "One in ten school-kids has asthma. When they're sick, their moms and dads worry. We need cleaner air to keep pollution out of our lungs."

What if Henry and Leona were in MY family? they wondered.

It was time to vote, and all hands rose for YES.
When the moment is right, progress can't be stopped.

Everyone celebrated when the new electric bus finally arrived.

Miss Leona sat behind the wheel. Instead of a roar, it purred and glided quietly onto the road.

A child pointed and waved, excited like young ones get about trucks and buses.

Miss Leona's stomach tightened, remembering when plumes of smoke covered people. But the air stayed clear.

Henry's words echoed in her mind:
science can help make things better.

Kids climbed aboard. “Wait, this bus is new!” a child shouted.

"It's better than new...it's electric," said Miss Leona. "And clean electricity can be made by wind, sun, and water."

Henry took a long, deep breath. "Today's ride is different. I don't feel like coughing. There's no smoke or nasty smell. Just fresh air."

"Maybe you won't need so much medicine," said his friend.

Henry took a marker and drew a new yellow bus on a blank card. “It’s a *Clear-Skies Ice Dragon*,” he said. “Miss Leona and I summoned it together.”

He gave her the card. “So you’ll always remember me, even on my sick days.”

"Thank you," she said. "For teaching me that we have to believe to make things real."

And just maybe, if we believe in a cleaner world, it will be a place fit for dragons.

Afterword

The left picture below was taken in New York City in 1900 – only one car can be seen among a sea of horses. Thirteen years later, the same street had transformed – only one horse among all the cars! Amazing how new technology can completely change things.

Today we're seeing a similar transformation with electric vehicles. Within a decade, the gas car may go from dominance to dinosaur. There are a lot of reasons to go electric: EVs are cleaner, quieter, safer, and can recapture energy to be more efficient. Some electric cars are so quick, it feels like a roller coaster when the driver pushes all the way on the pedal!

Since EVs have no tailpipe, they don't make smoke and fumes that lead to asthma attacks. Asthma can be passed down in families, and caused by pollutants in the environment like mold, tobacco smoke, and car exhaust. People can develop asthma due to breathing pollutants from nearby busy roads at home or work. During an asthma attack, the airways to the lungs become smaller, and the body makes mucus, blocking air from entering the lungs. For people with asthma, cleaning our air is welcome.

There's another big reason we all need EVs to replace vehicles powered by fossil fuels: the health of our planet, and survival of all the animals and plants we love and depend on. Climate change is causing extreme weather events, severe droughts, melting of the polar ice, and rising ocean levels. The pollution emitted by cars has played a big role. By switching to EVs that have no tailpipe emissions, and creating electricity from renewables like sun and wind, we can fight these catastrophic changes.

It will take determination and effort to transform our driving. Like Henry said – humans working together and using science can make a difference. But you have to believe...

Easter morning 1900: 5th Ave, New York City. Spot the automobile.

Source: US National Archives.

Easter morning 1913: 5th Ave, New York City. Spot the horse.

Source: George Grantham Bain Collection.

Made in the USA
Middletown, DE
30 October 2021